IMAGES
of America

SEATTLE'S
COAL LEGACY

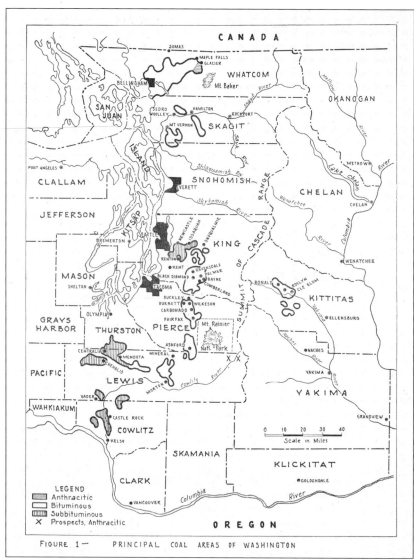

FIGURE 1— PRINCIPAL COAL AREAS OF WASHINGTON

This 1947 Washington State Division of Mines map shows the principle coalfields and coal towns of Western Washington. To the north, Bellingham was the first coalfield in Washington mined by the Hudson's Bay Co., in 1851. In 1859 and 1863, the Seattle area mines in Issaquah and New Castle were worked for the first time, followed by the Black Diamond mines in 1884. Tacoma's mines in Wilkeson and Roslyn were developed for the transcontinental Northern Pacific Railroad in the mid-1880s. The strip mine in Centralia Washington was the state's largest; it operated from 1970 to 2006. (Courtesy of Washington State Division of Mines.)

ON THE COVER: The Gilman Mine is pictured around 1890 in what would become Issaquah. Issaquah coal was first tested in Yesler's sawmill blacksmith shop in 1859. From the beginning on the Black River in 1854 up through 1974, when the last mine, Rogers No. 3 in Ravensdale, shut down, there were approximately a 110 operating coal mines and prospects in King County, home of Seattle. Port Blakely, an important shipbuilding center (Hall Bros.) and sawmill, is directly across Puget Sound from Seattle and is one of Seattle's suburbs reached by ferry. (Courtesy of University of Washington [UW] Special Collections, IND1465.)

IMAGES
of America

SEATTLE'S COAL LEGACY

John M. Goodfellow

ARCADIA
PUBLISHING

Published by Arcadia Publishing
Charleston, South Carolina

Printed in the United States of America

Library of Congress Control Number: 2019937714

For all general information, please contact Arcadia Publishing:
Telephone 843-853-2070
Fax 843-853-0044
E-mail sales@arcadiapublishing.com
For customer service and orders:
Toll-Free 1-888-313-2665

Visit us on the Internet at www.arcadiapublishing.com

CONTENTS

ACKNOWLEDGMENTS

I would like to acknowledge the photographers who contributed to this volume. As I started to research coal's impact on Seattle, it dawned on me that this period coincided with the portable camera and the development of photography when it was still a very chemical-based skill. The photographs were taken to show prospective investors, far away owners, or local interest the coal works. Photographs of mines, trains, and ships could be used for promotion and pride of ownership and reflect the spirit of the time. As a secondary piece of history, I had a great reliance on others. Much thanks goes to the local historical societies in New Castle, Renton, Issaquah, Roslyn, Wilkeson, and especially Black Diamond for keeping coal mining history alive. I especially want to thank Seattle's Museum of History and Industry (MOHAI), the University of Washington's Special Library (UWSL), and Puget Sound Maritime Historical Society (PSMHS) for curating most of these pictures.

I would also like to thank Joe Baar and Roger Ottenbach at Puget Sound Maritime Historical Society and Matthew Schmidt for their general assistance throughout the project.

INTRODUCTION

It is easy to forget that coal's steam-powered classical period only lasted 200 years, from 1720 to 1920. Seattle's major coal era was even shorter: 1870–1920. It is true that more coal is burned today, to power turbines for electricity, than ever before, but the days of coal-powered heating plants and steam engines in trains and ships as well as factories is gone. Today, coal competes with natural gas, nuclear, hydroelectric, and increasingly with alternatives. Though coal is "dirty," it is hard to imagine a human population that has easy access to coal, like India or the United States, to forego electricity generation from it.

Coal mining in King County lasted for over 100 years, from 1859 to 1975. The use of coal peaked during World War I, with many mines closing thereafter. The last major mine in King County, Rogers No. 3, closed in 1975. Up until 2006, coal was being strip-mined in Centralia, supplying 4.1 million tons to two steam turbines, but it was cheaper to import coal from Wyoming and Montana.

Coal is made by geologic pressure on vegetal matter with heat over time.

Peat, a soft coal, is compared to the harder bituminous and the hardest, anthracite. Coal here has about 9,600–10,000 BTUs per pound. Coal is easier to handle and burns hotter and longer than wood.

The use of coal has a long history; by the 2nd century AD, coal was being mined and used in Roman Britain. At that time, it was being exported by ship from the mines to London and the European coast. It was used to heat public baths and breweries and for ironwork. By the 16th century, with the denudation of forests near cities, coal was cheap enough to be used in home and industry. The first wooden tramroads were used to bring coal to water transportation routes for shipment to places that used the product. It is said that with the increasing use of coal and resultant soot that the white moths in London changed color evolutionarily to black, having to identify with the soot or be eaten due to the lack of camouflage.

The greater demand for ironwork and heat brought more need for coal. The greater need for iron was also causing the smelters to denude entire forests to make charcoal, which has less contaminants than coal, for smelting ore in Britain. In 1709, Abraham Darby developed an industrial plant that smelted ore using coke—cooked coal—on an industrial scale.

More supply begets more uses, and more uses mean more demand. The problem is that when one digs deep mines in wet climates, the water level is insurmountable without a mechanical pump. In the 18th century, that solution became a coal-fired, steam-powered engine attached to a pump. In 1712, Newcomen's Atmosphere engine condensed steam in the cylinder, creating a vacuum pulling in a piston. It could pump 150 gallons a minute from 160 feet down. This revolutionary machine was used in tin mines and coal mines in Britain, mostly in Wales and Cornwall. Quite a few still exist, and a few even do work.

In 1776, James Watt improved the efficiency of the steam engine by having the condensing go on outside the cylinder. This allowed the economical use of steam engines away from the coal mines. He also invented the steam chest, which used the expansive power of steam on alternating sides of the piston. This allowed for a variable rpm (revolutions per minute), and then, by using high-pressure boilers, it created many orders of magnitude of increased power. So many new uses were developed besides pumping out mines, such as powering trains, ships, and factories. The coal-fired industrial revolution had begun.

In the 19th century, water-powered turbines were developed; later, these were adapted to steam. Turbines are still used in nuclear power plants and coal plants with steam. Water-powered turbines connected to generators are used with dams to generate electricity.

During the early part of this industrial revolution, Seattle existed as an Indian village on a spit. The first steam engine on Puget Sound was the Hudson's Bay Company's steamship *Beaver*. It burned wood and had a three-horsepower Newcomen engine that used saltwater for the condensing. It operated from 1836 to 1888, when it hit the shore near Vancouver and broke apart.

After settling Alki in 1851, Arthur Denny opened a store to trade with settlers and passing ships coming and going from San Francisco gathering timber. He soon moved to a spit of land where Pioneer Square is today. At this point, ships from San Francisco were coming to Seattle to take timber to California where it had become scarce; it was being used to shore up mines and to build piers and houses. In 1852, Henry Yesler, who had previously built water-powered mills in Ohio, joined Denny and the other pioneers and set up the first steam-powered mill on Puget Sound; it included a blacksmith shop. Local pioneers and natives supplied the labor, and an industrial revolution in the wilderness had begun. When coal was found within 30 miles of Seattle, there already was an active trade with San Francisco.

During the Civil War, the Union Pacific Railroad crept across the country, connecting in Utah in 1869 with the Central Pacific Railroad eastbound from San Francisco. Once the transcontinental trains connected to San Francisco, ships served other West Coast ports, including Seattle. This created an enormous demand for coal.

The only coal in California was in Mount Diablo, 30 miles east of San Francisco. The mine was small, and the coal was not good quality. Soon, the entire town of Nortonville and the Black Diamond Coal Co. would move to King County. Seattle and its environs had lots of bituminous coal.

The San Francisco–based Central Pacific's contract with the New Castle mines across Lake Washington from Seattle made coal big business here, bringing capital and finance as well as important mining and mechanical knowledge and experience. Seattle, being well located as a regional and Pacific nexus with access to coal, soon made Seattle the center of Pacific Northwest culture and business.

By 1908, according to the US Bureau of Navigation, Puget Sound was the third-largest coaling port after Baltimore and New York. Eventually, with the development of California's oil industry, Seattle imported its petroleum energy from California and later Alaska.

Coal as a large, dominant industry only lasted 50 years here in Seattle, from 1870 to 1920. During that time, we made steam engines, boilers, and ships. With the experience gained in the coal age, Seattle went on to produce diesel engines and airplanes and later computer hardware and software. This is our industrial legacy. We got things done, and we do them still.

Electricity in Seattle was originally coal-powered. In 2019, Seattle City Light has 91 percent of its electric needs met from hydroelectric and one percent from coal. Wind farms on the other side of the mountains contribute four percent nuclear. While hydropower is the largest source of electricity here in the mountainous, wet Northwest, in most of the rest of the world, electricity is created by burning coal. Currently, coal produces 40 percent of the world's electricity and 70 percent of the world's steel.

One

DEVELOPMENT
1854–1900

Seattle was a village when the coal trade began and a boomtown thereafter. What began as a burlap bag trade advanced to trams and then railroads. Industrial needs were met in Seattle. By 1900, Seattle was producing mill and mining equipment, steam engines, ships, power plants, and other major technological products. By 1876, there were three blacksmiths, two boilermakers, and two foundries as well as three machine shops, including Seattle Coal and Transportation Co. (SC&T). Transportation was becoming more and more regular, connecting the communities of Puget Sound, regionally from Olympia to Victoria, to the coast to San Francisco, to South America, and to the Orient.

In the mid-1880s, Seattle was connected to Tacoma and Portland by the Northern Pacific Railroad and then later directly over the mountains to the Midwest and East Coast. Later, the Great Northern Railroad provided Seattle with another transcontinental railroad across Stevens Pass to the north. The railroads proved to be a boon for the timber industries and fisheries now having large markets to the east. The Great Northern also ran large regularly scheduled steamships to Asia, bringing silk from China and tea from Japan. Coming to Seattle were the products of the Midwest and East; merchandise, yes, but also settlers, grain, hides, meat, and other raw products.

By the late 1880s, there were five banks, two shipyards, one dry dock, five coal yards, four breweries, two creosote works (using coal tar), four railroad terminals, two gristmills, and headquarters for 160 steamboats. The Great Seattle Fire of 1889 destroyed most of the commercial district, but Seattle rebuilt better and stronger than ever.

When the gold rush hit in 1897, Seattle was prepared as a center of wholesaling, transportation, mechanical skills, and good public boosterism.

The end of the century saw new contenders in the energy market. In 1876, Well No. 9 in Pico Canyon, north of Los Angeles, became the first paying well on the West Coast. Electric generation plants were powered by coal, but 1899 saw the first hydroelectric facility on Snoqualmie Falls. Hydroelectricity and petroleum from California would eventually displace coal as the fuel for power.

Chief Seattle arranged labor for the first settlers to Seattle, led by Arthur Denny, below, in 1851. The first industry in the area was the procurement of timber for San Francisco. The timber was used in sawmills for lumber and piers and to support mine shafts and galleries. In 1855, the first load of timber was shipped to China. Arthur Denny and his brother David were involved in many of Seattle's early industrial projects, such as railroads, mines, and finance. Arthur was a territorial legislator nine times and worked for the charter and gave the land that the Territorial University would be built on. (Both, courtesy of MOHAI; left, SHS6; below, SHS3772.)

Henry Yesler was a millwright in Ohio before he moved to Seattle in 1852. Here, he assembled Puget Sound's first steam sawmill and metalworking shop, the beginning of Seattle's Machine Age. This blacksmith shop is where coal was first used locally. Both natives and pioneers worked there. It supplied the lumber for the first houses and stores and was traded to San Francisco, bringing money to the community. (Courtesy of UW Special Collections, PSE138.)

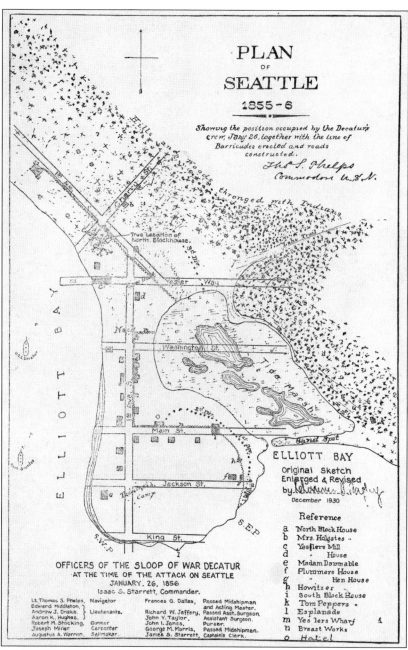

PLAN
OF
SEATTLE
1855-6

Showing the position occupied by the Decatur's crew, Jan.y 26, together with the line of Barricades erected and roads constructed.

Thos. S. Phelps
Commodore U.S.N.

OFFICERS OF THE SLOOP OF WAR DECATUR
AT THE TIME OF THE ATTACK ON SEATTLE
JANUARY. 26, 1856
Isaac S. Starrett, Commander.

Lt. Thomas S. Phelps,	Navigator	Frances G. Dallas,	Passed Midshipman and Acting Master.
Edward Middleton,	Lieutenants.	Richard W. Jeffery,	Passed Asst. Surgeon.
Andrew J. Drake,		John Y. Taylor,	Assistant Surgeon.
Aaron K. Hughes,		John I. Jones,	Purser.
Robert M. Stocking,	Gunner	George M. Morris,	Passed Midshipman.
Joseph Miller	Carpenter	James S. Starrett,	Captain's Clerk.
Augustus A. Warren,	Sailmaker		

Reference
a   North Block House
b   Mrs. Holgates "
c   Yeslers Mill
d   "   House
e   Madam Damnable
f   Plummers House
g   "   Hen House
h   Howitzer
i   South Block House
k   Tom Peppers "
l   Esplanade
m   Yeslers Wharf
n   Breast Works
o   Hotel

In 1853, Washington Territory was separated from Oregon, creating the state of Oregon. In 1854, about 300 tons of coal from Renton came down the Black and Duwamish Rivers and was loaded on the steamer *Major Tompkins,* which had the California to Puget Sound and Victoria, British Columbia, mail contract. At the beginning of the Battle of Seattle, three settlers were killed up the Duwamish, making coal deliveries untenable. This map shows the military situation during the Battle of Seattle in 1855–1856. Before the war, there were 170 nonindigenous Seattleites; after the war, 150. The map shows that the natives controlled the trail to Lake Washington, and the USS *Decatur,* a navy ship, is offshore. Madam Damnable's, the local inn and bordello, and Yesler's Mill and wharf are shown as well. (Courtesy of UW Special Collections, SEA1382.)

12

U.S. REVENUE CUTTER "LINCOLN."

In 1859, a sack of coal was brought from Issaquah and tested and used in Yesler's blacksmith shop. In 1863, more conveniently located coal, from what would become New Castle, was tested at the blacksmith shop and in the boilers of the US Revenue Cutter *Lincoln*, where it was found to burn "hot." Afterward, coal from New Castle would be brought across Lake Washington to a boat landing at Leschi and brought over the Lake Washington trail/road to Seattle by wagon. (Courtesy of Coast Guard Alaska.)

This picture from 1866 shows Yesler's original log cookhouse, built before the sawmill. It operated as a community center in Seattle's early years. In front of the building are some natives, who along with the pioneers, worked in the mill. Yesler's mill had a 12-horsepower steam-powered engine driving a 48-inch circular saw. The mill provided the locals with lumber with which to make framed planked houses like the office at left and building at right and brought money from export. (Courtesy of UW Special Collections, UW5870.)

The Washington Territorial Legislature incorporated Seattle in 1865, when this photograph was taken. The Territorial University, founded in 1861, can be seen on the hill. By 1865, three to five tons of coal a day were coming down the Black and Duwamish Rivers from New Castle. In 1867, when Alaska was purchased from the Russians, it stimulated trade, including coal. That same year, Thomas Martin built the first independent iron-working facility and foundry. (Courtesy of MOHAI, SHS239.)

This is a c. 1865 picture of Sarah Yesler, who was involved in many of Seattle's early civic organizations, including the Seattle Public Library. She was Seattle's first librarian. When she died in the 1880s, her husband, Henry, donated their house to the city as the first permanent library. In 1863, Seattle's first newspaper, the weekly *Seattle Gazette*, was published. It printed dispatches from the Civil War using the new telegraphic connection to Portland. (Courtesy of UW Special Collections, UWSHS6663h.)

Chen Chong and his wife are pictured here in 1866. Chen Chong opened a cigar shop and arranged Chinese labor, which was used extensively in building the western railroads, including those of Seattle. In 1886, after the Northern Pacific was completed, anti-Chinese riots took place that drove the majority of the Chinese out of the city. The year after this photograph was taken, Thomas Martin would open his iron and brass foundry. Seattle's population in 1866 was 550. (Courtesy of UW Special Collections, SHS4000.)

The steamship *Alida* was built in Seattle in 1870. She was built because the owners had secured the mail contract between Olympia, Tacoma, Seattle, and Victoria, British Columbia. This is also the year that Seattle's population of 1,107 surpassed that of Olympia. Olympia was the territorial capital and terminus of the road from Portland. Seattle was the hub of maritime transportation. (Courtesy of Seattle Public Library, SPL-SHP-5122.)

The year 1870 is when the Seattle Coal and Transportation Co. succeeded Lake Washington Coal, which owned the mines and transportation from the New Castle mines. Improvements were made using a wooden tram railroad, which took the coal to Lake Washington, then barged it to a landing where the Montlake Cut is now, portaged it into Lake Union, landed it at South Lake Union, and ran it up what is now Westlake Avenue, to the Pike Street coal bunkers and wharf. (Courtesy of UW Special Collections, UW36358.)

Henry Yesler and Chief Ouray are pictured here around 1870. Ouray, chief of the Utes of western Colorado, met with Presidents Lincoln, Grant, and Hayes, who called him "the most interesting man I've ever conversed with." Ouray went around the western United States warning other tribes of diminishing wildlife and sovereignty, as the natives, even on their reservations, were being deluged with settlers and miners. (Courtesy of MOHAI, SHS6271.)

In 1871, Seattle got its first locomotive, *Ant*, after the Seattle Coal and Transportation Co. was sold to San Francisco investors. The company had gotten a contract with the Central Pacific Railroad, which had connected in Utah with the Union Pacific as the first transcontinental railroad in America in 1869, for 5,000 tons of coal a month. The route was improved from New Castle to the lake and from South Lake Union to the bunkers and wharf at Pike Street with a steam locomotive. In 1871, before those improvements, 4,418 tons were mined; after improvements in 1876, a total of 104,556 tons were produced. (Courtesy of MOHAI, 2002.2.437.)

By 1875, regular connections to San Francisco were served by various vessels. The side-wheeler SS *Pacific*, seen here, was Seattle's first regular connection to San Francisco in the 1870s. The SS *Pacific* would, soon after this picture was taken, be struck by the sailing ship *Orpheus*, off Cape Flattery, with 273 dead and only 2 survivors. At right, the more modern SS *Salvador* ran the route from Panama to San Francisco until that trade died off due to transcontinental trains and then served Seattle. At right is a sailing schooner. (Courtesy of MOHAI, SHS11765.)

This picture from 1873 shows Yesler's mill down Mill Street, or what was known as Skid Row, where logs were slid down to the mill. The dark smoke could be caused by the use of coal to increase the steam pressure in the boilers when a lot of work is done. At right is a view looking west farther up Skid Row. Yesler's mill is in the distance. Where the flagpole stands is Pioneer Square, near the totem pole. In 1875, there were 3,100 white people, 250 Chinese, 50–100 Indians, and 300 transients in Seattle. In the same year, the steamer *Fanny Lake* and all her machinery, such as boilers and engine, were made in Seattle. By 1876, there were three blacksmiths, two foundries, two boilermakers, and two machine shops, one of which was that of Seattle Coal and Transportation Co. (Courtesy of MOHAI, SHS1108 and 2002.3.409.)

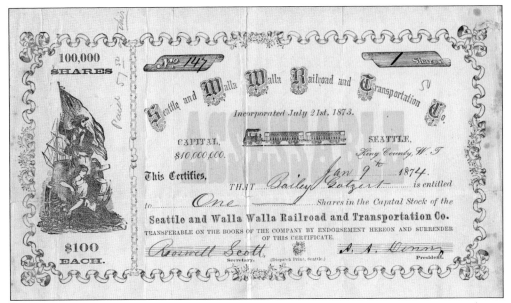

The Seattle & Walla Walla (SWW) Railroad was started as a citizen's response to Tacoma being chosen as the Northern Pacific Railroad terminal in 1873. On this Seattle & Walla Walla Railroad transportation stock certificate, issued to operating partner Baily Gatzert, of the wholesale house Schwabacher Bros. & Co., Arthur Denny signs as president. It was James Colman who financed, built, and ran the railroad between the Renton coal mines and new coal bunkers and wharf off King Street in 1874. In 1877, with Renton coal coming in, the coal bunkers on Pike Street collapsed; subsequently, the SWW Railroad was extended to New Castle in 1878. In 1877, a total of 400 to 800 tons of coal was reaching the King Street wharves a day. (Courtesy of UW Special Collections, PNW03746.)

James Murray Colman came to Seattle in 1872 to lease Yesler's mill after setting up the better-capitalized Port Madison Mill. A talented machinist and engineer, Colman built the Seattle & Walla Walla Railroad to the Renton coalfields after a citywide effort had failed. He also ran Colman Creosoting, which preserved pilings with coal tar; Colman Dock, the main mosquito fleet dock; and the extant Colman building across from it. (Courtesy of MOHAI, SHS88.)

The *George E. Starr*, named for the recently deceased founder of the Starr Line, was built in 1878 in Seattle for Puget Sound and Victoria regional service. Note the walking beam on top, used to transmit power from a cylinder to a crank on the side-wheeler. At this time, the King Street bunkers and wharf distributed coal from both Renton and New Castle on the Seattle & Walla Walla Railroad to colliers for San Francisco and other Pacific ports as well as to passenger and freight steamboats regularly operating locally, regionally, and to San Francisco and around the Pacific. She was one of the ships rushed into Yukon service at the beginning of the gold rush in 1897. (Courtesy of MOHAI, 1018-11.)

In this 1878 bird's-eye view of Seattle, before the Great Seattle Fire, both Pike Street and King Street coal bunkers and wharves and numerous coal stacks and steamboats can be seen. The train servicing the coal bunker on Pike Street can be seen at upper left, in the woods. In 1877, the Pike Street coal bunker and wharf collapsed. The King Street bunkers and wharf with coal

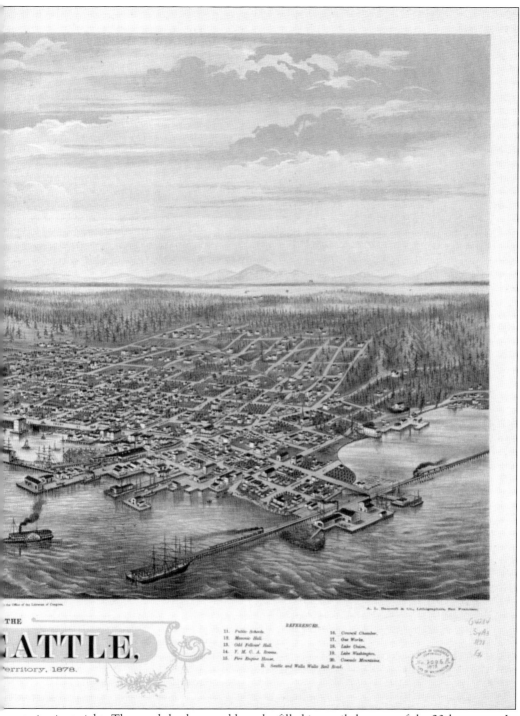

THE

ATTLE,

erritory, 1878.

trains is at right. The south harbor would not be filled in until the turn of the 20th century. In 1880, Seattle's population was 3,533; Olympia's, 1,240, Tacoma's, 1,103; and Portland's, 17,577. (Courtesy of Library of Congress.)

THE CITY OF SAN

BIRDS EYE VIEW FROM THE BAY LOOK

San Francisco, D. Mc Quillan, 605, Market St.—Agent for the P

NCISCO.

H-WEST.

This 1878 bird's-eye view of San Francisco, taken at the same time as the one of Seattle, shows a more urban city. While Seattle's population was 3,533 in 1880, San Francisco's was 233,959. The heart of the city was masonry, and there were numerous coal stacks south of Market Street and what looks like a soot cloud. San Francisco imported most of Seattle's coal, as its local coal, from Nortonville, east of the city, was of inferior quality and soon exhausted. The entire town of Nortonville was moved to Black Diamond in southeast King County in 1883. (Courtesy of Library of Congress.)

There were numerous small coal depots operating at the same time as the big ones. This picture shows the Talbot Coal Company of Renton's dock in 1878, situated on Seattle's waterfront. A high-wheeled delivery cart, for local distribution, is under the shed roof. Local distribution would be for heating plants and steam engines. Having local coal was an economic advantage to Seattle. (Courtesy of MOHAI, 1983.10.6141.)

In 1880, coal exports surpassed timber exports. The Talbot Coal Company depot can be seen at right, and the Territorial University is on the horizon. At Yesler's dock, a schooner collects lumber. Sailing ships transported commodities like lumber, coal, and grain up through World War I. In 1880, the Seattle Gas Light Co. started providing coal gas to the city's streetlights. (Courtesy of UW Special Collections, UW6996.)

Dexter Horton's bank developed when people started keeping their valuables at his store for protection. Pictured here around 1880 are Horton's partner A.A. Denny (third from left), Denny's son Rolland (far left), and perhaps two of Denny's grandchildren. Finding enough money to stay in Seattle was, and it seems will always be, of concern to those that live here. San Francisco and Eastern capitalists had the money needed for the large projects, such as mining and mills. Having a local bank was a great stimulus to business activity. In 1879, there were 11 colliers and three coal schooners that made port in Seattle. Besides the strong export market for coal by ship, there was also a large demand at home, for industry, transportation, and heating. (Courtesy of MOHAI, 1990.45.30.1.)

In 1881, James Colman and the investors of the Seattle & Walla Walla Railroad (SWW) sold their profitable coal railroad to Henry Villard. Villard, a German journalist during the Civil War, afterward worked for German railroad investors. Twice, he controlled the Northern Pacific Railroad. Here in Seattle, the SWW became the Columbia & Puget Sound Railroad and Oregon Improvement Co. When Villard bought the railroad and docks, he also bought Seattle Coal Co. in New Castle. (Courtesy of HistoryLink.org.)

The SS *Umatilla* was part of Villard's three-ship collier fleet—*Umatilla, Willamette,* and *Walla Walla*—that Villard had built in Chester, Pennsylvania, in 1881 for the Seattle coal trade. This picture was taken after 1888, when the SS *Umatilla* and the SS *Williamette* were converted to passenger liners for the San Francisco to Seattle route. (Courtesy of PSMHS.)

When Villard took over the coal railroad Seattle & Walla Walla, he acquired 2 locomotives, 50 coal cars, 2 passenger cars, a mail car, and 6 flat cars, all built in Seattle. He also improved the coal wharves and built a depot and a machine shop. To the right of the tracks and next to the water tower is Stetson and Post Co., a sawmill with log rafts in front. In the image below, Capt. John Williamson, of the side-wheel steamer tug *Yakima*, poses in front of the King Street coal wharf and bunkers looking north. (Both, courtesy of Seattle Public Library; above, SPL26436; below, SPL-SHP-5190.)

Here, the Columbia & Puget Sound Railroad brings in a shipment of logs from the east side of Lake Washington in 1880. Villard planned to connect with the Northern Pacific Railroad when it reached Tacoma from Portland and then across the continent from there. (Courtesy of MOHAI, SHS330.)

By the 1880s, logging machines, in the form of wood-powered steam donkey engines, were hauling trees through the woods to a railhead or a body of water; here, it was a river. Washington Iron Works of Seattle made some of the best machines. This picture, taken in 1903, shows the same basic technology used then. (Courtesy of MOHAI, 1983.10.6888.)

In the year 1884, when Washington women received the vote, the coal town of New Castle, across Lake Washington, was the second-largest city in King County. This was also the year the Columbia & Puget Sound Railroad reached Black Diamond and soon the Franklin, Taylor, Cumberland, Hobart, and Kumming coal mines. Nothing exists of this town now except one miner's house and some tumbledown trunk bunkers. (Courtesy of MOHAI, SHS3487.)

JUDKINS, PHOTO, COR. SECOND AND COLUMBIA STS., SEATTLE, W. T.

In 1884, Seattle was connected by train to Tacoma and Portland and hence across the continent. At this time, Seattle was Puget Sound's center of finance, wholesale, transportation, and manufacturing. This view from the top of the King Street bunkers is looking at the Oregon Improvement Co.'s Ocean and City docks and warehouses, connecting Northern Pacific rail to Seattle's maritime transportation. Railroad officer James McNaught has been quoted as saying, "There's one car to Tacoma for every ten to Seattle." Below, a steam train with a load of Northern Pacific boxcars sits next to the Columbia & Puget Sound depot. (Both, courtesy of MOHAI; above, 2008.54.23; below, SHS11262.)

Rail workers are at the Green River crossing of the Cascades branch of the Northern Pacific Railroad over Stampede Pass. The North Pacific Railroad, after connecting Seattle to Tacoma and Portland, began its more direct route over the Cascade Mountains. In 1885, about 15,000 workers (out of 20,000) on the line were Chinese, like the man with a scarf. In 1886, Seattle's anti-Chinese mobs ran a majority of the Chinese out of town. That year, Edison's Seattle Electric Co. provided the first incandescent lighting system (DC) on the West Coast. Its DC generators were coal-powered. (Courtesy of MOHAI, SHS5907.)

In 1887, the Northern Pacific Railroad would cross Stampede Pass, and in 1888, workers would complete the tunnel. To cross the Rockies, the Great Plains, and then the Cascade Range is a daunting task that takes organization and large amounts of capital. This snow-removal gang and trestle bridge show two important elements for such a crossing, but fuel, tracks, bridges, depots, water sources, salespeople, dispatchers, and trains were all needed. (Courtesy of MOHAI, SHS5910.)

Seattle's Yesler Street cable car ran up the old Indian trail to Lake Washington at what is now Leschi in the 1880s. There, it connected with steamers going to the east side of the lake and Mercer Island as well as coal barges from New Castle and Renton. Coal barges would have supplied energy to the steam engines in Leschi that pulled the cable around. Passengers and freight, including coal, were pulled by cable over the hill to Seattle. (Courtesy of MOHAI, SHS723.)

The ferry *City of Seattle* ran between West Seattle and Marion Street in downtown Seattle from 1888 to 1907. It was operated by the West Seattle Land Improvement Co., which sold land there. It was the first double-ended ferry on Puget Sound and carried carriages, horses, and people. Her boiler was 5.5 feet in diameter and 22 feet long; she ran two 135-horsepower engines at 100 psi and made the two-mile run in eight minutes. (Courtesy of PSMHS.)

In 1889, the *City of Pueblo* was on the Seattle–Victoria–San Francisco route when she made a record run of 50 hours from Victoria, British Columbia, to San Francisco. At 326.5 feet long with a 38.5-foot beam, she was pretty typical of the coastal ships of the time. (Courtesy of UW Special Collections, HES050.)

A sailing collier is docked at the end of a King Street bunker around the time of Great Seattle Fire on June 6, 1889. That year, Williamson & Kellog built eight steam engines, logging railcars, sawmill machinery, and 15 hop furnaces. The railroads were proving an economic stimulus as commodities in hops, timber, and fish moved east. (Courtesy of UW Special Collections, SEA 2610.)

The picture above, taken before the Great Seattle Fire, looks down Commercial Street, now First Avenue, toward Mill Street. Streetlights and sidewalks are present, but the road is still dirt. The entire commercial center and railroad facilities would burn down in the fire, but not the docks. Below is a picture of the city center at the corner of Mill and Front Streets. After the fire, Commercial Street and Front Street were joined where Pioneer Square is today, and Mill Street became Yesler Way. Note the soldier standing guard in the ruins. (Both, courtesy of MOHAI; above, 1983.10.6343; below, SHS2340.)

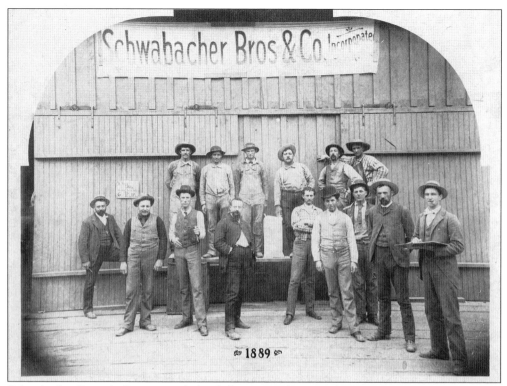

≈ 1889 ≈

Schwabacher Bros. & Co. was a wholesale house supplying work camps and small communities with everything from coffee, tobacco, and sewing supplies to anchors and hardware. Its dock and warehouse were not damaged in the fire. Above, Schwabacher's shows it is still in business after the fire. Below, Dexter Horton's bank proclaims its existence with a picture of its safe in the doorway. Soon, Seattle will be rebuilt of brick and stone. In 1889, Washington became a state. (Both, courtesy of MOHAI; above, 1981.7279.1; below, SHS2385.)

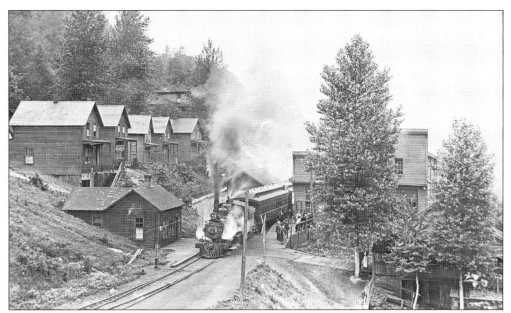

The Franklin Mine was located east of Black Diamond, along the Green River Gorge, and opened the year after Black Diamond was founded, in 1885. Soon, it employed 450 men and produced 800 tons a day. The mine closed in 1919; today, it exists only as archaeological remains. (Courtesy of MOHAI, SHS6050.)

In 1891, the Knights of Labor called for a strike at the Oregon Improvement Co.'s mines in New Castle, Gilman (Issaquah), Black Diamond, and Franklin. The governor called in the National Guard when riots broke out between Pinkerton guards, an armed miner group, and the black miners who were brought in, unbeknownst to them, as strikebreakers and found themselves in a violent situation. Two men were killed. (Courtesy of MOHAI, SHS2022.)

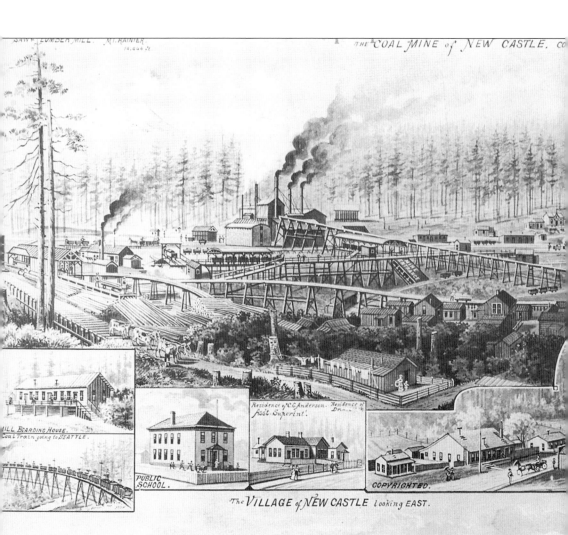

SAW & LUMBER MILL. MT. RAINIER. 14,444 ft. THE COAL MINE of NEW CASTLE. CO

MILL BOARDING HOUSE. Coal Train going to SEATTLE.

PUBLIC SCHOOL.

Residence of C.C.Anderson. Residence of Asst. Superint. Dru...

COPYRIGHTED.

The VILLAGE of NEW CASTLE looking EAST.

J. SMITH, Gen'l. Manager.

C.C.ANDERSON, Assist. Superint.

B. COREY, Superint. of New Castle & Franklin Coal Mines.

VIEW of NEW CASTLE, CO near SEATTLE, WASHINGTON.

40

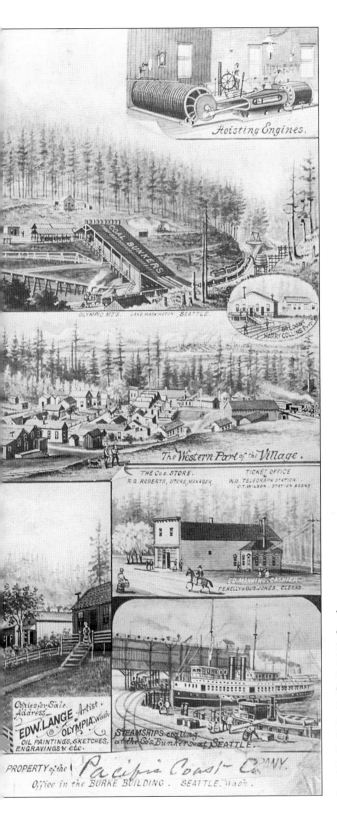

This drawing is a bird's-eye view of Coal Creek and New Castle, the coal works, and the village. In the 1890s, New Castle was the second-largest city in King County. The entrance to the mine is seen at center left, with a steep, sloped trestle leading down. The vignettes include a steam-powered hoist in the upper right corner and the King Street bunkers at bottom right. (Courtesy of Newcastle Historical Society.)

Steamer service operated on Lake Washington first from Leschi and then in 1892 from Madison Park; there were coal-handling facilities at both. Transporting coal first from New Castle and then Renton was economical by barge. In 1885, a lock at Montlake Cut could take barges and small ships between Lake Washington and Lake Union. In this c. 1891 picture is the ferry *Kirkland*, which connected with the Yesler cable car at Leschi. Ferries from Leschi and Madison Park served east-side towns Juanita, Kirkland, Houghton, and Mercer Island. (Courtesy of UW Special Collections, LAR057.)

After the Great Seattle Fire in 1889, many sawmills, shake mills, and shipyards moved to Salmon Bay in Ballard. The stern-wheeler *Baily Gatzert*, shown here, was built in the Ballard shipyard of John Holland with machinery from Pittsburgh. She was a fast steamer on the Seattle to Olympia route. (Courtesy of MOHAI, SHS1076.)

The *Wanderer* was considered one of the best towboats on the sound. Built in 1890, across the sound from Seattle in Port Blakely at the Hall Bros. shipyard, she was part of the fleet that moved log rafts around and brought sailing ships from the mouth of the Strait of Juan de Fuca to Puget Sound ports to load coal, lumber, wheat, and other commodities until after World War I. Williamette Iron Works of Portland made the steam engine, but the boiler was made in Seattle. A tugboat is mostly engine. The *Wanderer* had a 150-horsepower steam engine and a 10-foot propeller. (Courtesy of UW Special Collections, UW3416.)

By 1891, when this view was sketched, Seattle had been rebuilt after the Great Seattle Fire. The population had gone from 3,533 in 1880 to 42,837 in 1890. The Pike Street coal bunkers were gone, but the King Street bunkers had an additional wharf. At left is the Great Northern Railroad dock with rails laid at Smith Cove. There are quite a few more coal stacks in the city since the

1878 bird's-eye view on pages 22–23 was taken. The southern tide flats are still not filled in. The Rainier Brewery building with coal stacks and railroad access is at far right. (Courtesy of Library of Congress.)

The Seattle, Lake Shore & Eastern Railroad (SLS&E) was created in 1888. Here, its terminal and depot are seen at the bottom of Madison Street around 1893. It first traveled along the north shore of Lake Union, then to the northwest shore of Lake Washington and the east side of Lake Sammamish, and finally to the coal mines in Issaquah. Now, the railbed is called the Burke Gilman Trail, after the railroad's two founders. After reaching the coal mines, the railroad connected with the Canadian Pacific in Sumac, Washington, giving Seattle a second transcontinental connection. After this, the SLS&E was bought by the Northern Pacific Railroad. (Courtesy of MOHAI, SHS5020.)

Here, Great Northern Railroad workers lay the last rail in Stevens Pass on December 10, 1892, giving Seattle its third transcontinental connection. The Great Northern's service to St. Paul and points east in 1896 connected with steamships at its terminal in Seattle's Smith Cove, providing regular service via Japan's mail line, Nippon Yusen Kaisya (NYK), to Yokohama, Japan. The biggest imports were silk and tea. In 1910, an avalanche came down close to this area, wiping two trains off their tracks and killing 96 people. (Courtesy of MOHAI, SHS11033.)

Peter Kirk, of Scotland, built Moss Bay Iron Works in what became Kirkland in anticipation of connection to the SLS&E's line to Snoqualmie Pass, where Arthur and David Denny had an iron ore mine that turned out to be more disappointing than anticipated. When the Economic Panic of 1893 happened, all development ceased. There were two blast furnaces and a foundry. (Courtesy of MOHAI, 1983.10.6113.)

This is a picture of the coal bunkers in Tacoma in 1893. When the Northern Pacific Railroad came to Tacoma from Portland in 1884, the Pierce County coalfields at Wilkeson were tapped. With the Cascade branch over Stampede Pass in 1887, coal mines in Roslyn and Cle-Elum were developed. In 1892, an explosion in its Roslyn mine killed 45 miners. The Grain Terminal is the building in the distance. (Courtesy of UW Special Collections, WAT008.)

Arthur Denny owned the Denny Clay Co. Mine in Taylor, east of Maple Valley, in 1895. Later, he was succeeded by his son Orion, pictured above at left with John Prichard, the mine superintendent. The Taylor mine and factory produced sewer pipe and brick and had its own coal mines. It probably produced the sewer pipe below around 1895, when Seattle completed its first waste and storm drain system. In 1895, about 118,000 tons of coal was exported and a significant amount used locally. A metal coal stack can be seen at left below. (Both, courtesy of MOHAI; above, 1962.2540.5; below, Lib1991.5.78.)

The US Revenue Cutter *Golden Gate* was the first steel vessel built on Puget Sound. It was built by the Moran Brothers Co. in 1897 and used by US Customs, with customs workers meeting and boarding vessels in San Francisco. Below, Robert Moran poses next to a propeller and rudder of a ship at his shipyard south of the King Street bunkers around 1899. In 1898, *Mike Maru*, part of the Japanese mail line (NYK), began regular service to Japan and the Orient, meeting trains in Smith Cove. Robert Moran was mayor from 1888 and 1890, during which the Great Seattle Fire happened. (Above, courtesy of PSMHS, 10526; below, courtesy of MOHAI, 1995.49.124.)

In 1897, this crowd at the Oregon Improvement Co. docks says goodbye to the old collier *Willamette*, recently refitted for passengers and headed for Alaska and the goldfields in the Klondike. Seattle supplied 75 percent of the ships and most of the merchandise destined for Alaska—and still does today. Before the gold rush, 18 ships plied the Alaska trade; after the gold rush, 173 did. (Courtesy of PSMHS, 1650.)

The Moran Brothers Co. built these 12 stern-wheelers for Yukon River service. There were 24 first-class cabins and room for 200 if there was no freight. The two engines generated 700 horsepower. In 1898, Robert Moran accompanied them to Alaska, where one sunk along the way. They towed barges of merchandise such as candles and kerosene from Standard Oil of California. Coal was gathered at exposed coal seams along the shore, though wood was probably the more common fuel. (Courtesy of MOHAI, 1988.33.217.)

Coal Bunkers
at Seattle Wash.
Poto. By L. HEATH. July. 28th 1898

During the 1896 Depression, Villard's local enterprises went broke. In 1897, the Pacific Coast Co. bought Villard's Oregon Improvement Co., the Columbia & Puget Sound Railroad, and the coal mines in New Castle, Issaquah, Black Diamond, and Franklin as well as improvements such as machine shops, depots, and the King Street coal bunkers and wharf pictured here. (Courtesy of UW Special Collections, IND0340.)

At the end of the 19th century in 1899, Seattle got its first hydroelectric plant, at Snoqualmie Falls. Shown are the four initial 1.5-megawatt generators in an underground room at the bottom of the falls. The power was sent to Seattle, where there were two existing coal-powered generator plants. (Courtesy of UW Special Collections, WWDL0467.)

Two

MATURATION
1900–1920

As the 20th century began, Seattle was a steam-powered city with many kinds of factory machinery; heating plants; electricity; and coal gas. It was also a steamship center for local, regional, and transpacific freight and passenger liners and three soon-to-be four transcontinental railroads. All of it was coal-fired.

Seattle was now a major industrial center, building ships of both wood and steel and producing boilers and steam engines as well as mining, mill, fishing, and timber machinery. In 1909, the city and the Pacific Northwest showed the world what it had accomplished at the Alaska-Yukon-Pacific Exhibition.

Petroleum from California had begun to seep into coal's dominance as an energy source, but local coal use continued shooting up, and exports peaked during World War I, after which petroleum took over, and mines began to shut down. Hydroelectric power was also of increasing importance during this time as well, though, like petroleum, it did not come to replace coal until after the war. It takes time for the infrastructure of prospecting, mining, processing, storage, and distribution to reach a level where replacement becomes possible.

During World War I, 20 percent of ships were built here. After the increased production of World War I came the General Strike of 1919. Wage regulations that had been put into place during the war ended, and wages were going down. The shipyard workers got sympathy from the other local unions, which at that time were more industry based like caterers and trolley workers. There was little work on the horizon, and the owners feared that a strike might precede a workers' revolution, like the recent Bolshevik Communist Revolution in Russia in 1917. They also worried when "Big" Bill Hayward, head of the Industrial Workers of the World (IWW) union, advocated for a worker-dominated political system. A total of 950 troops, 600 police, and 2,400 special deputies met the strikers. The five-day strike was successful in that it showed what could be done tactically if not effectively.

Gasoline, a distillate of petroleum, allowed new lightweight engines that were used in cars and planes, the high-tech gadgets of their time. Seattle's industrialism could evolve beyond coal because of the knowledge and experience already in place in the city and county.

DRILLING BOILER PLATE

This Moran Brothers Co. employee drills holes in a boilerplate. A boiler must have very exact rivet holes. This sheet will be turned into a cylinder, riveted together, and then used under high pressure. The Moran Brothers Co. made not only ships and boilers, but also machinery, pumps, and engines. In 1900, there were 12 shipyards employing 1,400 men and 34 foundries. (Courtesy of MOHAI, 2000.87.49.)

Intake and Buildings from North Side.

By 1901, when this picture was taken, the hydroelectric plant on Snoqualmie Falls was producing 13.7 megawatts of power. The pyramid-shaped structure was the original elevator cover leading to the underground turbines. The brick building was the transformer room. In 1901, there were 953 manufacturers in Seattle and 381 in Tacoma. (Courtesy of UW Special Collections, WAS1671.)

Compare this 1900 picture of the Issaquah Mine to the 1890 Gilman Mine on the cover, and contemplate the difference in technology. The wood stacked on the right was used in the mines as support. The large mines in New Castle and Black Diamond had their own sawmills and electric power plants powered by coal. (Courtesy of MOHAI, 1978.6585.53b.)

In 1869, a log flume was dug at the Montlake Cut to bring logs from Lake Washington to the sawmills at the south end of Lake Union and Fremont. This lock, built around 1900, allowed barges and boats hauling coal from New Castle and Renton to move through. The ship canal to Puget Sound would not be built until 1916. (Courtesy of MOHAI, Lib1991.5.14.)

This is a c. 1900 picture of the Bayview Brewery, which would become the Rainier Brewery, now west of Interstate 5. Coal was used for boiling; producing steam to run engines; and for cleaning, bottling, and canning. To the left, steam comes out of the powerhouse, and at right are railroad tracks, bringing the required ingredients, including hops from Auburn and later eastern Washington. Coal was also used by the author's great-grandfather to can and evaporate milk, which was an important export, especially to Alaska. (Courtesy of MOHAI, SHS12871.)

The Centennial Mill Co. was on the south end of Commercial Street, near Morgans's Shipyard, around 1885. Grain from eastern Washington was brought by train to Seattle and Tacoma for processing and export. By this time, waster mills had been supplanted by steam mills for grinding grain and other things, such as sawing wood. This is a 1905 picture of the mill with the powerhouse on the right; note the metal coal stack. (Courtesy of MOHAI, SHS10211.)

Carl Wallin, a shoemaker, met John Nordstrom in the Yukon. In 1901, they formed a partnership and opened a shoe store in Seattle. At this point, Seattle was the center of Northwestern merchantiling, finance, and transportation. Wallin & Nordstrom Shoe Store would become Nordstrom Best in 1929. (Courtesy of MOHAI, 1986.5.12080.1.)

Seattle's population was 80,671 in 1900. There were 21 civil engineers, 13 mining engineers, 7 mechanical engineers, and 4 electrical engineers, of whom Walter Gordon Clark was one. His company, Kilbourne and Clark (1899), made radiotelegraph equipment for oceangoing ships and had clients like the Australian Post Office. The equipment was also used by ships on the West Coast. (Courtesy of MOHAI, 2013.111.1.)

These are pictures of the mine in Franklin, east of Black Diamond, in 1902. This location is where the mine cars came out of the mine. They would proceed around the track to the bunkers on the left leading to the picking tables pictured below, where the coal was cleaned of rocks and bone and graded by size. In 1902, the state's output of coal exceeded a million tons—1,102,211. (Both, courtesy of Washington State Historical Society; above, 1943.42.1052; below, 1943.42.1058.)

Having transcontinental railroad connections meant that canned and iced fish could reach Midwestern and Eastern markets. The mechanization of the fishing industry is shown below in the gutting and de-finning machine, invented in Seattle. In 1901, a total of 1,380,590 cases of salmon were canned on Puget Sound. By the early 1900s, Puget Sound's catch eclipsed that in the Columbia River. Soon, Seattle was sending fishing boats to Alaska and basing canneries there. (Both, courtesy of MOHAI; above, SHS10593; below, SHS16233.)

The Hall Bros. shipyard moved from Port Ludlow in 1879 to Port Blakely across from Seattle on Bainbridge Island. Henry Hall, right, ran the place until his son James, left, took over in 1903. Henry's brother Winslow designed the tug *Wanderer*. Winslow, on Bainbridge Island, is named for him. James created the carved scrollwork found on the bow of the *Wanderer*. This picture of father and son was taken around 1903. (Courtesy of MOHAI, SHS16557.)

Port Blakely, where the Hall Bros. shipyard was located in 1902, was also home of the Port Blakely Mill, built in 1864. The sawmill could produce 100 million board feet a year, with the trees coming from around Puget Sound. The lumber was loaded on sailing ships bound for California, Hawaii, South America, and Australia. The sawmill provided the lumber for the shipyard next door. (Courtesy of MOHAI, 1983.10.18160.1.)

In 1903, James Hall moved his shipyard to Eagle Harbor, where the Washington State Ferry system has its docks and shops now. This 1904 photograph from the Hall Bros. shipyard shows three wooden ships in three stages of construction. From 1874 to 1903, the Hall Bros. built 108 wooden ships, both sail and steam. (Courtesy of MOHAI, SHS1146.)

In 1901, Moran Brothers Co. got the contract to build the Virginia-class battleship USS *Nebraska*. Here in the foundry, a high-pressure cylinder head is cast. Below can be seen the Moran Brothers Co.'s machine shop, neat and tidy, the day of the laying of the *Nebraska*'s keel on July 4, 1902. Note the men and overhead crane for scale. (Both, courtesy of MOHAI; above, 1988.33.249; below, 1995.49.274.2.)

Here, the USS *Nebraska*'s enormous pair of 19,000-horsepower, triple-expansion steam engines are to be fed from 12 coal-fired boilers. The Moran Brothers Co. made all the machines in house. Using the ghost of a man at lower right as a reference makes the sign seem proportionally two by six feet and the engines 25 feet high, approximately. Some kind of smaller engine sits on the floor. During the world cruise of 1908, the USS *Nebraska* received the "E" award for efficiency of coal use. (Courtesy of MOHAI, 1995.49.39.)

Launched in 1904, the USS *Nebraska* joined the Great White Fleet in 1908, cruising around the world and visiting Pacific ports in San Francisco, Hawaii, New Zealand, Australia, Philippines, Japan, and Ceylon. The fleet was called "white" because of the peacetime white paint on its hulls. Along the way, sources of coal had to be bought and stockpiled. (Courtesy of www.history.navy.mil.)

In this undated photograph are, from left to right, Frank, Robert, and Sherman Moran of Moran Brothers Co. In the late 19th century, the Moran Brothers Co. produced many ships, including the first all-steel ship built on Puget Sound. At the beginning of the 20th century, the company produced the battleship USS *Nebraska,* making all the machinery, such as the boilers, engines, and pumps. The organizational skills, to say nothing of the mechanical, was extraordinary. Robert, the head of the organization, retired soon thereafter to Rosario, his estate, the land of which became Moran State Park on Orcas Island in the San Juan Islands. (Courtesy of MOHAI, SHS1688.)

Vulcan Iron Works started out as Seattle Iron Works in 1883, the year they built eight large steamboat engines. Vulcan and Moran Brothers Co. were two of the many machine shops and foundries building engines and boilers for steamships, sawmill machinery, pumps, and heating plants in the coal era. The many machinists who worked in these shops were a highly skilled workforce. (Courtesy of UW Special Collections, CUR 781.)

Seattleites were constantly griping about the crudeness of this Northern Pacific depot on Railroad Avenue along the waterfront. The Seattle Electric Plant in the background on Post Alley was built in 1902 and still exists. The picture below is looking over the waterfront. Both images were captured around 1905. (Both, courtesy of MOHAI; above, 1983.10.6939; below, 1983.10.7093.1.)

In 1901, William Pigott had an industrial supply house. Later, in 1903, he founded the Seattle Steel Co., below, which forged pig iron from scrap. They also rolled out merchant shapes like C-channels, squares, rounds, and rebar. This plant still exists doing the same work under the West Seattle Bridge. In 1905, he founded the Seattle Car Manufacturing Co., which built logging railroad cars and coal cars. Later, this company moved to Renton and eventually became PACCAR, building train cars, military tanks, logging trucks, and highway trucks. (Both, courtesy of MOHAI; above, SHS17801; below, 2000.107.165.22.02.)

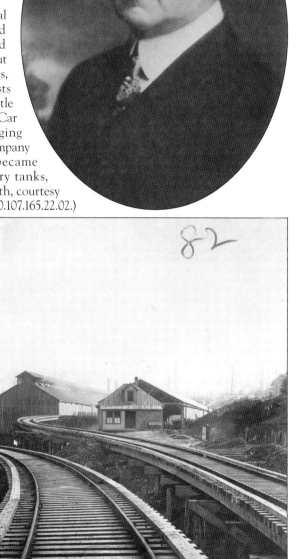

In this c. 1905 photograph, the *Dakota* and *Minnesota*, owned by Jim Hill's Great Northern Railroad, lie at the Great Northern docks at the north end of Elliot Bay in Smith Cove. A Great Northern Railroad express train, the *Oriental Limited*, met the ships and hurried silk to the East Coast silk factories. (Courtesy of MOHAI, SHS208.)

When the *Minnesota* and *Dakota* were built in 1904 at the Eastern Shipbuilding Co. in New London, Connecticut, they were the largest freighters constructed in the United States at that time. Soon, because of their slow speed, they became known as "Hill's White Elephants." They provided regular service from Great Northern Railroad docks at Smith Cove, along with their eight Japanese Mail Line (NYK) fleet mates, to Asia. By this time, wireless telegraph was becoming common. (Courtesy of PSMHS, 1606-6.)

The *Dakota* was too big to fit in any shipyards in Seattle and was taken across the sound to the Puget Sound Navy Yard dry dock in Bremerton for work. The Moran Brothers Co. built the powerhouse and pumps that empty the dry dock in 1895. The Navy requirement was for a pump and engine that could move 200,000 gallons a minute. (Courtesy of PSMHS, 696-7.)

After the Great Seattle Fire, the wharves with warehouses were rebuilt at an angle to the shore to accommodate trains. Cargo came from all over the Pacific. These longshoremen unloaded boxes of shredded coconut into a warehouse; later, they will load them onto a train that will speed them east. In 1901, China Mutual Steam Navigation Co. had direct service scheduled from Seattle to Mamilla, Calcutta, Hong Kong, and Singapore. (Courtesy of MOHAI, SHS7490.)

This is a 1909 picture of Pacific Coast Coal Co. workers on electric mules and coal cars near New Castle at Coal Creek. Pacific Coast Coal Co. had taken over Villard's interest during the 1896 Depression. In 1910, three-and-a-half million tons of coal would be produced statewide, a threefold increase in 10 years. The major coal mines of the area generated their own electric power. (Courtesy of UW Special Collections, CUR1694.)

This small steam yard tug would have switched railcars in a small yard, like a lumberyard. Wood was still used to produce steam where it was convenient and economical, like within lumberyards and sawmills. (Courtesy of MOHAI, 2014.59.58.)

In this 1906 picture, two steam tugboats tow the sailing ship *Ganges*, of Norway, into Puget Sound to load lumber at Port Blakely, a Seattle suburb. The ship was bound for Callao, Peru. Sailing ships were important in the movement of commodities like coal, grain, and lumber through World War I. (Courtesy of MOHAI, 1983.10.PA7.26.)

In 1906, the steam-powered electric turbine plant in Georgetown was built. Inside were General Electric Curtiss turbines, only invented 10 years before and powered by coal-generated steam. Such turbines were found to be more efficient than reciprocating engines and are still used today in coal and nuclear plants to generate electricity from the expansive force of steam. This picture was taken in 1985; the plant is still extant. (Courtesy of MOHAI, 2000.107.173.07.02.)

The Seattle Lighting Gas plant came to North Lake Union in 1906, relocated from eastern Pioneer Square, because of a plentitude of fresh water and access to coal by barge and railroad. By cooking coal, volatile coal gas is released; it is similar to natural gas. Coke, the by-product, is used for smelting ore. Here, coal bunkers are seen. This plant became today's Gas Works Park. (Courtesy of MOHAI, 1983.10.10656.3.)

In 1906, Seattle finally got the railroad terminal it thought it deserved: King Street Station. The Great Northern built it but shared it with the Northern Pacific. The clock tower, imitating the one in Venice on St. Marks Square, had beautiful tiled waiting rooms inside. (Courtesy of MOHAI, 1974.5868.445.)

The steamship *West Seattle* was built in 1907 in Tacoma by the train yard to supplement the steamship *City of Seattle* on the West Seattle run for the West Seattle Land Improvement Co. It met a cable car in West Seattle that went to the Admiral district. (Courtesy of UW Special Collections, TRA 780.)

When her six-year-old child Willis died, Anna Herr Clise and 23 female friends organized what became Seattle Children's Hospital. Her husband, James Clise, owned part of and managed the Globe Navigation Co., which operated steamers and sailing ships around Puget Sound and the entire Pacific. (Courtesy of MOHAI, SHS18884.)

As can be seen from this 1909 picture of delivery wagons, Seattle's domestic market for coal was strong. Many old Seattle houses have coal chutes leading to the basement furnace room. The mining of coal locally gave Seattle industry the advantage of cheaper energy costs. (Courtesy of UW Special Collections, SEA 1789.)

By 1892, there was 50 miles of streetcar track and 22 miles of cable car track. In 1902, an interurban reached Kent. As seen here in 1910, a trolley that ran from downtown to Ballard connected with the Ballard-to-Everett interurban seen here. Both freight and passengers were carried on the line. Its railbed is now used in north Seattle as a trail park. (Courtesy of MOHAI, SHS735.)

Pictured above are engines and boilers in the foundry building at the Alaska-Yukon-Pacific Exhibition 1909. Donkey engines, used in logging, can be seen at left. In this exhibition, Seattle showed the world and itself what was happening in this once-remote corner of the earth. There were ethnological exhibits as well as girly shows. (Courtesy of UW Special Collections, UW28514z & MOHAI, 1974.5868.168.)

The year 1910 saw the completion of washing down Denny Hill, expanding the commercial district northward, and also the filling of the Duwamish tide flats, creating Harbor Island. Reginald H. Thomson was chief city engineer off and on from 1892 to 1931, during which time he supervised large city projects, such as the work seen here. (Courtesy of MOHAI, 1983.10.8164.)

Pumps, powered by steam, drain water out of the Ford mine in the Pacific Coast Coal Co. town of Coal Creek, upstream from New Castle. Machinery requires special knowledge and training to operate and maintain. (Courtesy of UW Special Collections, CUR623.)

Coal mining before World War I was big business. Puget Sound was the third-largest coal port in the United States, with 3.5 million tons produced in 1910. This picture shows the mine head and bunkers of the Issaquah Mine in 1910. During the war, the Issaquah Mine was shut down, as a German national owned it. The mine below was in Renton at the same time as the one above. (Both, courtesy of UW Special Collections; above, WAS 0180; below, MOHAI, SHS17516.)

The smelter in Irondale, near Port Townsend, first opened in 1901. It was located there because of easy access to trees to make charcoal, which, due to its purity and heat, was used for smelting iron. Charcoal does not impart impurities to making iron. By 1910, when this picture was taken, coke (cooked coal) from Pierce County replaced coal. Note the smoke coming out of the stacks that has been drawn in for an industrial look. (Courtesy of MOHAI, SHS8436A&B.)

In 1919, the Brace and Hergert mill, located in South Lake Union where MOHAI is now, drew its logs from the east side of Lake Washington through a log flume before the small lock where the Montlake Cut is, before the ship canal was built. Since coal could be brought this way, it would not be surprising if it was used to give a boost to the boilers, as seen by what appears to be coal smoke. (Courtesy of MOHAI, SHS9881.)

By 1910, the Denny Clay Co. had become the Denny-Renton Clay and Coal Co. The picture above is of the Taylor Clay Works, which produced brick and sewer pipe. Below, the Van Asselt plant in Georgetown is alongside the railroad lines south of the city. Coal was used to fire the clay and power mechanical machines. (Both, courtesy of UW Special Collections; above, CUR788; below, CUR786.)

The photographs above and below show two sides of the plant at Renton. Clay and coal were found near here. In the picture above, clay and coal go into the factory; below, products go out by train. Terra-cotta architectural details were produced here that grace many old Seattle buildings, such as the Cobb, Arctic Building, and others on the University of Washington campus. (Both, courtesy of UW Special Collections; above, CUR789; below, CUR790.)

Around 1910, there were 19 major mines operating in King County. The Renton Mine's slag heap that would later be used as fill for Boeing's Renton Field is pictured here. The design of the slag lifts were modified to create ski lifts. Around this time, a toboggan lift was built in Truckee, California. (Courtesy of UW Special Collections, WAS0865.)

In 1897, when the Pacific Coast Co. took over Villard's Oregon Improvement Co., the Columbia & Puget Sound Railroad, and subsequently the important mines in the region, it also modernized the coal bunker and wharves at King Street with conveyor belts, shown here in 1910. A loaded coal car is on the left while a sailing collier is at right. The log rafts are for the Stetson and Post Co. mill next door. Water access allowed these logs to come from anywhere around the sound to be easily transported, stored, sorted, and then processed by workers using machinery to process them. (Courtesy of MOHAI, 1983.10.6872.1.)

A Mr. Currin (in the foreground) stands with employees and product at his shoe factory in Seattle. From an early time, cattle were driven over Snoqualmie Pass, from eastern Washington, to be processed here in Seattle; later, they came by train. Steam engines were used to power all kinds of machines used in manufacturing. In 1910, Seattle's population was 237,194. (Courtesy of MOHAI, 2013.114.1.)

While this 1910 picture of steam retorts was taken at the Pacific American Fisheries plant in Bellingham, the same technology would have been used here in Seattle for cooking and sealing cans of salmon. Coal was used not only to can salmon, evaporate and can milk, brew and bottle beer, and heat large institutions like schools and hospitals, but it was also used to run steam laundries in South Lake Union, where there is still a coal stack. (Courtesy of MOHAI, 1983.10.8342.)

Here, the employees of the Coolidge Propeller Co. pose next to the bronze blades of a pitch adjustable propeller in 1910. Leigh Coolidge was a respected naval architect who designed/built propellers for ships such as freighters and who later designed high-speed propellers for hydroplanes. (Courtesy of MOHAI, 2005.44.1.)

In 1911, the Seattle Dry Dock Co. built a series of F-class submarines that tested to 200 feet deep. In 1913, using the patents of the Electric Boat Co. of Groton, Connecticut, the Seattle Dry Dock Co. built H-class submarines; the Chilean Navy ordered two of these submarines. These two were subsequently sold to British Columbia at the start of World War I. This is the steering room of H-3 *Garfish*. A chain drive leads from the helm to a shaft controlling the rudder. (Courtesy of navsource.org.)

Three nuns, who, in 1878, had the contract for the King County Poor Farm in Georgetown, started Providence Hospital. Industrialization of health care meant new skills and technologies requiring training and certification. In 1907, a nursing school was started. This c. 1911 picture shows the hospital nurses on the front steps of the hospital, which was to the east from where the Central Library is now. (Courtesy of MOHAI, SHS7583.)

Seattle now had not only the beautiful King Street Station (right), but, in 1911, the elegant Oregon-Washington Station as well. The Milwaukee Railroad and the Union Pacific Railroad served the new station. It was in 1911 that the Milwaukie Railroad came to Seattle after crossing Snoqualmie Pass. The Milwaukee Railroad took the path that Interstate 90 takes now and is the John Wayne State Park Trail, which crosses over the mountains and plains of Washington. (Courtesy of MOHAI, 1997.19.3.)

Above, in 1913, blacksmiths make metal work for the construction of the Smith Tower, seen at left. For many years, it was the tallest building west of the Mississippi River. The building is steel framed and covered in terra-cotta tiles. The triangular building is the Hotel Seattle, built in 1890. When it was torn down in the late 1960s, it motivated Seattle's history buffs to cherish and preserve Pioneer Square as a historic district. A parking structure stands there now. (Both, courtesy of MOHAI; above, 1983.10.9738; left, SHS15393.)

The South Lake Union City Light plant was built in 1912. At the time, coal would have come by barge from Lake Washington and the railroad. In this picture, taken in the 1920s, a sailing vessel turned into a coal barge is at the dock in front. Directly next to the plant is a hydroelectric plant, which got its water from the overflow of the Volunteer Park drinking-water reservoir. To the right of that is a coal-powered manufacturing plant. Below is a coal barge around 1912 selling retail coal at the wharf. The sign reads, "Buy Coal Now, Monks Miller Coal and Wood." (Both, courtesy of MOHAI; above, 1983.10.608; below, 2002.3.1819.)

On July 18, 1913, Navy sailors, after a previous street brawl, smashed up the International Workers of the World (IWW) headquarters and burned its books. The IWW advocated militant revolutionary socialism, wishing to create a society controlled by the workers. In 1916, the boat carrying them to Everett in support of the shingle weavers strike got in a gun fight with the sheriff; 7–12 were dead, 47 wounded. (Courtesy of MOHAI, 1983.10.7800.)

Industrial work could be dark and cold, and accidents were frequent. Here, coal pickers clean rocks and bones from coal around 1913 at the Issaquah Mine. At this time, the United Mine Workers of America had organized the mine workers. (Courtesy of MOHAI, 1978.6585.53d.)

This is a picture of the old Seattle Steel Plant now owned by Pacific Coast Steel Co. in 1914. In Europe, war had begun, and it would be an industrial war. The plant is the one that exists now below the West Seattle overpass. To the left, at the end of Harbor Island, is Todd Shipyard. The machines roll out hot metal to various shapes and thicknesses. (Both, courtesy of MOHAI; above, CUR1108.)

While there were quite a few large mines, there were many more small ones. Here, a narrow-gauge coal train railway of the Snoqualmie Coal and Coke Co. comes out of the woods. (Courtesy of MOHAI, 1983.10.9569.)

Hyde Coal Co. of Cumberland is a good example of a medium-sized mine in King County in 1914. The boilerhouse in the foreground, with the metal coal stack, is fed water from a wooden aqueduct. Farther back are the mine slope and bunkers. Where the steam comes out is where the steam engine and hoist is that pulls the coal and men out of the mine. (Courtesy of UW Special Collections, CUR1065.)

Pictured here are the Hyde Coal Co. steam engines that lift the coal cars and men up and down the tilted coal seams. Bell signals and marks on the cable guided the operator where to stop at the appropriate locations of the various galleries, which extended horizontally from the shaft. From the galleries, the coal was mined and then brought to coal cars, which then returned to the surface. Once on the surface, the coal was taken to the bunkers for cleaning and sorting. (Both, courtesy of UW Special Collections; above, CUR1069; right, CUR1718.)

Pictured here are two views of the Pacific Coast Coal Co.'s Black Diamond mine No. 11, also known as "the Morgan Slope," in 1915. At the time, this was the deepest coal mine in North America. It was 6,000 feet long and 2,200 feet vertical with a slope of 25 degrees; the top of the slope is seen here. The wood in the foreground of the image above was used as mine supports. The railroad cars below the bunkers are destined for Seattle. (Both, courtesy of UW Special Collections; above, UW23734; below, UW23730.)

Next to mine No. 11 was Mine B, pictured here. The next five years, with World War I, would see peak coal production in King County. During and after the war, petroleum from California and hydroelectricity became more available and convenient. Black Diamond stopped mining and sold the town after the Morgan Slope closed in 1927. (Courtesy of MOHAI, 1978.6585.38.)

On a West Seattle beach sits a seaplane (1910) about seven years after the Wright brothers' historic 1903 flight. This aircraft has essentially the same technology as today: light structural strength, light useful skin, steerage with a wheel and rudder, and a lightweight power source. (Courtesy of MOHAI, 1983.10.6517.2.)

At the beginning of the automobile's history, many cities had factories producing different versions of what a motorcar is. This Ajax model was designed, manufactured, and sold by the Parker brothers, Charlie, Frank, and George. Above is a picture of their 1914 six-cylinder, 70-horsepower engine and auto frame; below is their gearbox. This same year, Evenrude Outboard Motors were used on fishing dories, replacing the "ash breeze." (Both, courtesy of MOHAI; above, 1995.78.2; below, 1995.78.1.)

Large companies like Pacific Coast Coal Co. and Northern Pacific Railroad owned all the major mines in the area, like the Coal Creek Mine, seen above around 1915. There were many small mines, though, below the smallish Cannon Mine in the Green River Gorge. This type of mine is called a "water level" mine because no pumps are needed, as the coal seams slope out the entrance, which drains the water. (Above, courtesy of UW Special Collections, IND0430; below, courtesy of MOHAI, 1978.6585.43.)

Ravensdale, a town and mine in eastern King County, had only been recently sold to the Northern Pacific Railroad when an explosion killed 31 workers in 1915. The mine never reopened. What once was a company town is now a small hamlet (population 1,000) in the Western Cascade foothills. Near Ravensdale is where the last mine to shut down in King County was located; it was Rogers No. 3 in 1974. (Courtesy of UW Special Collections, WAS0861.)

A man works a coal seam in Issaquah around 1915. A drill bores holes for dynamite that will collapse the seam and be gathered onto railcars. The opening of the Panama Canal that year proved a boom for local coal producers. After the major mines closed in the 1920s and 1930s, they leased them to gyppos, or truck miners who mined the easy-to-get coal pillars. Truck miners supplied institutions like University of Washington, Western Hospital, and Puget Sound Navy Shipyard until the 1980s. (Courtesy of MOHAI, 1978.6585.53e.)

Built in 1909, the new Sisters of Providence Hospital on Second Hill, now Swedish Cherry Hill, is seen here around 1915. The coal stack is still there. Historic hospitals as well as schools, jails, libraries, and apartment and office buildings all used coal, and their stacks are monuments to the coal era. (Courtesy of MOHAI, 1983.10.10006.)

Shown here is the transformer house of the Seattle City Light hydroelectric plant in the Cedar River watershed on May 4, 1915. The power plant and dam were built at the instigation of Reginald H. Thomson, who was Seattle's long-serving city engineer (from 1892 to 1931). He supported the public ownership of utilities. He organized Seattle City Light and had a hydroelectric plant put in the city's watershed in 1911. (Courtesy of MOHAI, SHS17481.)

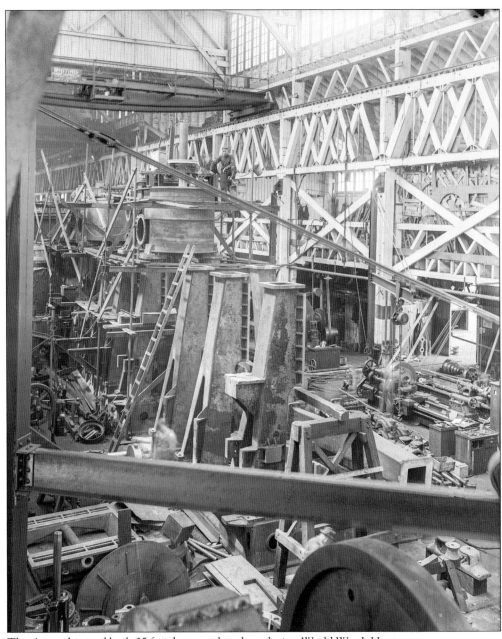

The Ames shipyard built 25 freighters and tankers during World War I. Here, an enormous steam engine is worked on by four men—one near the bottom of the ladder, one on a cylinder head, one on the machine at right, and one to the left of the ship bit. World War I was the impetus for the technological shift from coal to oil. This steam engine's boiler would be fired by oil, not coal. (Courtesy of PSMHS, 10016-10.)

The tug *S.L. Dowell* was purchased along with the Madison Park coal yard below by the Pacific Coast Coal Co. prior to the ship canal completion in 1916. Moving coal and logs by tug was cheap and efficient. The *S.L. Dowell* would sink off Leschi in a storm. Her bell was recovered and is at the Center for Wooden Boats. (Both, courtesy of MOHAI; above, 1978.6585.21d; below, 1978.6585.21b.)

Before the locks at the head of Salmon Bay on Puget Sound were completed, travel between Lake Washington and Lake Union through locks at Montlake Cut was carried out. With the completion of these new locks in 1916, a ship canal gave access from Puget Sound through Lake Union to Lake Washington. Many shipyards and other maritime industries sprung up along the banks of the canal and in Lake Union. (Courtesy of MOHAI, 2002.3.2022.)

Shingle weaving was dangerous work. The man in the back operates a powered crosscut saw cutting logs to length. The ones in the front cut the blocks into shingles. Before and after the Great Seattle Fire, Ballard was a center of the lumber industry. Lumber was sent on to consumers by both ship and train. After the canal was built in 1916, the mills modernized, becoming even bigger. (Courtesy of MOHAI, 1983.10.PA18.77.)

By 1915, petroleum from California was making big inroads on coal by making new uses like personal transportation possible. Above is a Shell Oil Co. retail station on Eastlake Avenue. Below is the Standard Oil distribution center on Harbor Island. Standard Oil of California, the successor to Pacific Oil Co., had opened Seattle's first gas station in 1907 at the dock. (Both, courtesy of MOHAI; above, SHS17057; below, 1983.10.10307.)

In the 1916 image above is Boeing's first airplane, the B&W, at the company's seaplane hanger on Lake Union. Below is the Martin seaplane that the company used for comparison purposes, with a rotary engine—perhaps it was an experiment. The coal gasworks can be seen in the distance. In 1917, William Boeing started the aeronautical program at the University of Washington and had the historic wind tunnel built there. (Both, courtesy of MOHAI; above, SHS11540; below, SHS9306.)

Pictured at right is a portrait of William Maltby, the superintendent of the Pacific Coast Coal Co. in 1916. At the time, Pacific Coast Coal had seven major coal mines in operation as well as railroad, maritime, and other infrastructure. In 1917, just prior to the US entrance into World War I, the Federal Fuel Administration was set up by the federal government to control fuel prices. Pictured below is the coal company's office and the dispatcher's room for the Columbia & Puget Sound Railroad. (Right, courtesy of UW Special Collections, POR1027; below, courtesy of MOHAI, 1978.6585.12a.)

Pictured is a shipyard worker using a pneumatic riveting gun in 1917. This is the same year as the Bolshevik revolution in Russia. Directly after the war in early 1918, the General Strike happened. Over 36,000 workers shut down the city in support of the shipyard workers, whose wages were being decreased as production decreased. Below is the Skinner and Eddy Shipyard, successor to the Moran Brothers Co., during boom war times; a coastal steamer is in front, and a battleship is in the distance. (Both, courtesy of MOHAI; left, 1983.10.10621; below, 1983.10.PA5.20.)

David Rodgers, who had overseen the building of the USS *Nebraska* for Moran Brothers Co., is seen here as manager of the subsequent Skinner and Eddy Shipyard in 1918. He oversaw the building of 75 vessels during the war; this shipyard completed more ships than any other in Seattle. (Courtesy of MOHAI, 2009.56.156.)

On June 15, 1917, the Fremont Bridge over the ship canal opened. The metal bridge cantilevers so ships can pass through. Before there was a trestle bridge connecting Westlake Avenue with Stone Way. This year saw the Bailey Gatzert turned into a car ferry using elevators for the Seattle–Breamerton run; it was the first car ferry to cross the sound. The Seattle Mets hockey team also won the Stanley Cup against the Montreal Canadians in 1917. (Courtesy of MOHAI, 2006.7.2.)

During World War I, Puget Sound Bridge and Dredging Co., founded in 1899, built wooden ships. In 1917, Seattle built 44 wooden ships and 72 steel ships and employed 17,000 people, making the city a major shipbuilding center during World War I. (Courtesy of MOHAI, 1995.37.19.2.5.)

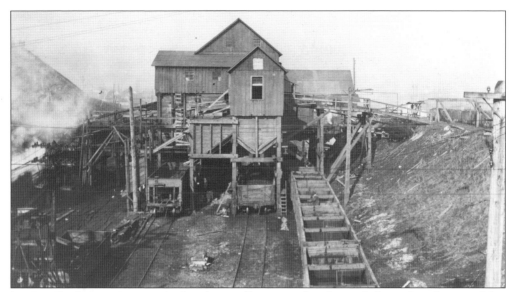

In 1918, Washington's mines produced 4.1 million tons of coal, a peak for coal production. After World War I, petroleum and hydroelectric begin to dominate. The diminished use of coal combined with the labor strife from 1919 to 1923 shut down more and more mines. Here, a Renton mine appears to be thriving in 1919. (Courtesy of UW Special Collections, IND0302.)

This is a picture of the launching of the steamer *Snoqualmie* on August 11, 1919. Puget Sound Bridge and Dredging Co., on Seattle's Harbor Island, launched this Ted Geary design, which was the biggest wooden vessel built to this time. Geary would design some of the area's most distinguished yachts, such as *Malibu*, *Blue Peter*, and *Canim*, as well as commercial boats and tugs. (Courtesy of MOHAI, 1995.37.19.2.46.)

Puget Sound Sheet Metal Works, founded in 1902, showcases ship supplies, such as these ventilators and lifeboats, during World War I. Later, the company would become a maker of airplane parts for Boeing. Today known as PSF Industries, the company designs and makes boilers, storage tanks, stacks, and other items fabricated from steel. (Both, courtesy of MOHAI; above, 2004.27.1.11; below, 2004.27.1.13.)

Three

DEMISE AND LEGACY
1920–1954

In 1920, there were 40,000 driver licenses issued in the city of Seattle. Petroleum and hydroelectric were making great inroads into coal as a source of power. Coal use and mining peaked during World War I and declined thereafter. Regardless, coal was being mined locally until the 1970s, although most mines shut down during the Depression and after World War II. Ravensdale was closed after an explosion killed 31 in 1915. Issaquah mines owned by a German national closed during World War I and never reopened. New Castle/Coal Creek stopped mining in 1929 after a fire destroyed the bunkers. Still, gyppo miners, independents or truck miners, continued mining the coal pillars until 1963. Coal mining in the Black Diamond district was still going on through World War II. The last mine to close in King County was Rogers No. 3 mine in Ravensdale in 1974.

During World War II, industry in Seattle and its suburbs stepped up. Tanks were built at what would become PACCAR, which had previously been producing train cars. Boeing famously produced masses of B-17 Flying Fortresses. Ships were built and repaired here. Fort Lawton, where Discovery Park is now, served as an embarkation depot during World War II, with a million soldiers passing through to the Pacific theater.

After World War II, petroleum and hydroelectricity dominated the local energy scene. Coal's legacy was not only the skills and know how that continued, but some of the physical assets as well. The Seattle Steel plant in West Seattle continues to produce merchant shapes from scrap, now using electricity. New Castle/Coal Creek and Squak (Issaquah) coal beds became Cougar Mountain Park. The Seattle, Lake Shore & Eastern Railway tracks that brought coal from Issaquah became the Burke-Gilman and Lake Sammamish bike and pedestrian trails. Alongside the Burke-Gilman trail in Seattle, the old coal gasworks became Gas Works Park. The old railroad bed from Black Diamond to Renton mines, along the Cedar River through the Maple Valley, is also a trail. Numerous coal stacks such as those at the University of Washington, Providence Hospital, Swedish Cherry Hill, Paramount Theater, and many others still stand as witnesses to the local coal era.

Coal's dominance has come to an end. Though more coal is burned now, it is all used to generate electricity. There is still a lot of local coal in the ground. To mine it though is too expensive, with cheaper coal coming from Montana and Wyoming to the steam turbines. Petroleum has only been used since 1900 and dominant for a 100 years, from 1920 to 2020—will it be replaced by cleaner alternative sources, as coal both was and was not?

In 1921, Boeing received a contract to build 200 MB-3A fighter planes. William Boeing, a trained engineer, had bought a boat shop on the Duwamish River, where his yacht *Taconite* was built, and it became his aircraft factory, known as the "Red Barn." Boats and airplanes at this time are both wooden and share many similar technologies such as rudders, shrouds, skin, and structure. (Courtesy of MOHAI, SHS9274c.)

Seattle's population in 1920 was 315,312. The longshoremen seen here are loading dry goods from a warehouse into a boxcar. There is a dock on the other side. Seattle was a transshipment point to and from points around the Pacific, especially Alaska. By 1922, there was a public cold storage in Smith Cove for storing fish and fruits for transshipment. Now, longshoremen use huge forklift-like trucks and cranes to move shipping containers between trains and ships The Longshoreman's Union has remained relatively strong while other industrial unions struggle. (Courtesy of MOHAI, SHS1221.)

Puget Sound Machinery Depot's boiler shop is pictured around 1922. The company, founded in 1887, worked on engines and boilers for maritime, railroad, and sawmill machinery. Here, various sizes of shaped boilers are assembled. In 1923, a total of 409 intercoastal vessels and 252 transpacific vessels visited the port. They went to the following places: Europe, 117; South America, 48; and Hawaii, 28. Grain and lumber were the biggest exports. Local Puget Sound traffic would have been higher still. (Courtesy of UW Special Collections, IND0285.)

By the 1920s, decentralized coal use was in demise, with many mines closing in the next 10 years, being replaced by cheaper, cleaner petroleum and hydroelectric power. Here, Seattle City Light's gorge power plant, on the Skagit River, can be seen. In the year 2000, this dam and powerhouse supplied 20 percent of Seattle's electrical needs. Electricity, which can be used to power, heat, and illuminate on demand, proved more economical than coal. (Courtesy of MOHAI, SHS17483.)

By 1925, extensive infrastructure had been completed by the Seattle city engineers: streets, sidewalks, water, storm drains, sewers, electricity, coal gas, and bridges. Pictured here is the Department of Roads and Sewers concrete transportation car on a sewer below a street. (Courtesy of MOHAI, 1968.4524.3.)

R.H Thomson, city engineer from 1891 to 1931, is pictured in his retirement year of 1931. He transformed Seattle from a town to a city, overseeing city infrastructural projects such as streets, sidewalks, bridges, water, sewer, and Seattle City Light. The filled-in Duwamish River tide flats are at center, and the ship canal is at left. (Courtesy of MOHAI, 1986.5.43455.)

Here, a seaplane refuels at the Standard Oil Co. dock on Harbor Island. By 1925, petroleum and electric infrastructure were in place to supplant coal's use. Below in 1926 is Baker Dam, on the Baker River, owned by the successor to the Snoqualmie Fall's plant, Puget Sound Power and Light Co. It produces 79 megawatts of electricity and serves areas outside Seattle, which is served by dams up the Skagit River. (Both, courtesy of MOHAI; above, 1983.10.12100.1; below, 1983.10.3286.1.)

Pictured here around 1927 is coal miner Ted Rouse at the New Black Diamond Mine in Maple Valley. The miners were mostly immigrants from Eastern Europe and the Ottoman Empire. By 1929, the Pacific Coast Coal Co. had closed all its mines but its modern New Black Diamond one. The town of Coal Creek was subsequently dismantled by 1932 and lots in Black Diamond sold off. Today, only a few traces of these mines exist. (Courtesy of MOHAI, 1983.10.6474.)

The New Black Diamond Mine in Maple Valley, also known as "Indian Mine," was state of the art when it opened in 1926, but challenging geologic conditions and a decreasing market closed it in 1941. At its height, it employed 300 miners and produced over 2.4 million tons of coal. (Both, courtesy of MOHAI; above, 1983.10.647.1; below, 1983.10.647.2.)

The *Quilicene* was rebuilt on Lake Washington at Houghton on the east side in 1929–1930 to accommodate cars on the Edmonds–Port Townsend run. She retained her 1916 triple-expansion steam engines from when she was called the *Kitsap II*. The new car ferry could carry 36 automobiles and 517 passengers. The interior was trimmed in mahogany, and there was a dining room. Her 1,200-horsepower steam engine moved her along at 14 nautical miles per hour. (Courtesy of UW Special Collections, TRA573.)

MCKA-IZING — Our Own process of pre-conditioning. Your new *Ford* Car "running in" with Coal Gas.

From 1906 until 1937, the gasworks on Lake Union supplied coal gas to the city and coke to smelters. Coal gas was used for everything that natural gas would be, including powering automobiles as seen here in 1929. A gas main went north to Mukilteo and south to Renton and Kent. After 1937, the plant was converted to petroleum supplied by barge through the ship canal from saltwater. In 1956, natural gas from Canada closed the plant. In 1962, the City of Seattle purchased the abandoned grounds for a park. (Above, courtesy of UW Special Collections, SEA6740; below, courtesy of MOHAI, 1983.10.176.94.)

Seattle City Light's Diablo Dam on the Skagit River was started in 1927 and completed in 1930. At 389 feet, it was the tallest dam in the world and could produce 129 megawatts. The Ross Dam, built in the 1940s, is higher up on the Skagit River and is named after James Ross, a successor to city engineer R.H. Thomson. Today, more than 90 percent of Seattle's electric power is hydroelectric. (Courtesy of MOHAI, SHS17484.)

Pictured here is Standard Oil Co. of California's Point Wells tank farm in 1932. It received its first shipment of benzine in 1912 from Southeast Asia. With access to maritime transportation and trains, this giant distribution center south of Edmonds, in Richmond Beach, served the local regions' increasing appetite for petroleum. It was originally an asphalt plant. (Courtesy of MOHAI, 1983.10.18037.)

Blanchard Boat Works was one of many boat yards on Lake Union. It built both pleasure and work/ military boats. Here in 1937 can be seen the yacht *Tolo*. Borrowing ideas from aircraft designs, boats became lighter and more streamlined. Below is a group of Blanchard's Senior Knockabouts, a popular, fast boat built on a production line to keep expenses low. Many Blanchard boats are used and loved today. (Both, courtesy of MOHAI; above, 1989.89.109; below, 1989.89.60.)

In 1935, before World War II, Boeing developed the XB-17, the prototype for the famous B-17 Flying Fortress bombers. The specifications required a plane that could carry a 2,000-pound bomb load 2,000 miles at 200 miles an hour. By November 1940, a year before Pearl Harbor, when the photograph below was taken, 7,500 employees were employed 24 hours a day. (Both, courtesy of MOHAI; above, 1986.5.155; below, PI23926.)

By 1943, local industries were working overtime developing war material. Before the war, Pacific Car and Foundry built railcars and logging trucks; during the war, it built Sherman tanks on an assembly line. Below, workers at Isaacson Steel Works machine shop lathe propeller shafts. John Isaacson had started a blacksmith shop in Seattle in 1907. (Both, courtesy of MOHAI; above, 1986.5.6759.7; below, 1986.5.4446.1.)

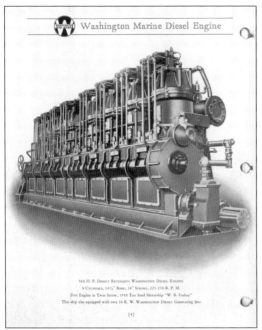

In 1883, Washington Iron Works employed 50 men; cast 1,000 tons of iron in its foundry; and produced three steam-powered mill engines, eight marine steam engines, and one locomotive. Washington Iron Works transitioned in the 1920s to produce large diesel engines that replaced steam engines. For instance, the Arthur Foss, now at South Lake Union Park, got its steam engine swapped out for a Washington diesel in 1938. (Courtesy of PSMHS.)

Puget Sound Sheet Metal Works employees fabricated B-17 and B-29 assemblies during World War II. More than 310,000 women worked in the aircraft industry, making up 65 percent of the workforce during the war, which was up from one percent prewar. (Courtesy of MOHAI, 2004.27.1.47.)

Olson and Winge Marine Works, originally founded in 1899, is seen here on the ship canal in 1944. The war brought a huge increase in the production, repair, and conversion of all kinds of ships and boats. Pictured above are, from left to right, machinists Frank Smith, Homer Pricket, and Axel Olson. Below, patrol rescue boats are at the Olson and Winge Marine Works yard. (Both, courtesy of MOHAI; above, 1972.5503.135; below, 1972.5503.109a.)

During World War II, with petroleum in short supply, coal had a last gasp of relevance. Seen here in 1943, a showroom promotes the domestic use of coal using modern techniques, such as "automatic coal stokers." Nearly all the remainder of the mines ceased after World War II and turned over leases to truck miners who mined the galleries and easy-to-reach coal pillars. They supplied large institutions like the University of Washington and the Western Mental Hospital and the Puget Sound Navy Yard. Up in the mountains, the Roslyn mine, owned by the Northern Pacific Railroad, was only open intermittently. These miners look pleased to receive their paychecks in 1955. Shown are, from left to right, Toby Wakkuri, unit foreman; Joe Sample, mine electrician; and Alec Bell miner; they all lived in Cle-Elum. (Both, courtesy of MOHAI; above, 1983.10.14671; below, 1986.5.5608.)

The last mine to close in King County was Rogers No. 3 in Ravensdale, in 1974. Coal's legacy is the skills and knowledge this community fostered during the coal/steam era that were applied to the age of petroleum and electricity, computer hardware, and computer software. Here, on May 4, 1954, Boeing rolls out the prototype to the 707, the Dash 80, onto the tarmac of its Renton airfield, which had been constructed on slag from the Renton coal mines. This is the same plane that barrel rolled over Sea-Fair. (Courtesy of MOHAI, 1986.5.188.)

The other legacy of coal is recreation. In the 1970s, Gas Works Park was created on the grounds of Seattle Gas Works on Lake Union. The old Seattle, Lake Shore & Eastern Railroad bed, which used to bring coal from Issaquah coal mines, was converted to a bike/pedestrian trail. It was named Burke-Gilman in honor of the railroad's founders. In the 1980s, the coalfields under Cougar Mountain became a county park. Other rails-to-trails coal roads exist in Maple Valley and upper Pierce County. (Courtesy of MOHAI, 1986.5.53567.)

DISCOVER THOUSANDS OF LOCAL HISTORY BOOKS
FEATURING MILLIONS OF VINTAGE IMAGES

Arcadia Publishing, the leading local history publisher in the United States, is committed to making history accessible and meaningful through publishing books that celebrate and preserve the heritage of America's people and places.

Find more books like this at
www.arcadiapublishing.com

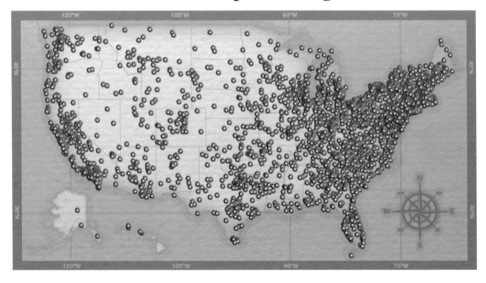

Search for your hometown history, your old stomping grounds, and even your favorite sports team.

Consistent with our mission to preserve history on a local level, this book was printed in South Carolina on American-made paper and manufactured entirely in the United States. Products carrying the accredited Forest Stewardship Council (FSC) label are printed on 100 percent FSC-certified paper.

MADE IN THE USA